This book b

If found, please contact me at:

mithmoth

Made in United States
Troutdale, OR
01/15/2024

16931499R00076